CAMBRIDGE LIBRARY COLLECTION

Books of enduring scholarly value

Earth Sciences

In the nineteenth century, geology emerged as a distinct academic discipline. It pointed the way towards the theory of evolution, as scientists including Gideon Mantell, Adam Sedgwick, Charles Lyell and Roderick Murchison began to use the evidence of minerals, rock formations and fossils to demonstrate that the earth was older by millions of years than the conventional, Bible-based wisdom had supposed. They argued convincingly that the climate, flora and fauna of the distant past could be deduced from geological evidence. Volcanic activity, the formation of mountains, and the action of glaciers and rivers, tides and ocean currents also became better understood. This series includes landmark publications by pioneers of the modern earth sciences, who advanced the scientific understanding of our planet and the processes by which it is constantly re-shaped.

Essay on the Modifications of Clouds

An industrial chemist by profession, Luke Howard (1772–1864) proposed the method of cloud classification that is still in use today. His life-long interest in meteorology led him to produce this landmark work in the history of the subject. General scientific opinion at the time was that clouds were too changeable to be classified, but, inspired by Linnaeus' work in biological classification, Howard proposed a method which used Latin terminology – cirrus, cumulus, stratus and nimbus – to provide a standard description for each of three groups of cloud types. His work was first published in the *Philosophical Magazine* in 1803; it was produced in book form in 1832 but went quickly out of print. This reissue is of the third printing (1865) of the edition brought out after his death in 1864 by two of his sons. Howard's other meteorological works are also reissued in this series.

Cambridge University Press has long been a pioneer in the reissuing of out-of-print titles from its own backlist, producing digital reprints of books that are still sought after by scholars and students but could not be reprinted economically using traditional technology. The Cambridge Library Collection extends this activity to a wider range of books which are still of importance to researchers and professionals, either for the source material they contain, or as landmarks in the history of their academic discipline.

Drawing from the world-renowned collections in the Cambridge University Library, and guided by the advice of experts in each subject area, Cambridge University Press is using state-of-the-art scanning machines in its own Printing House to capture the content of each book selected for inclusion. The files are processed to give a consistently clear, crisp image, and the books finished to the high quality standard for which the Press is recognised around the world. The latest print-on-demand technology ensures that the books will remain available indefinitely, and that orders for single or multiple copies can quickly be supplied.

The Cambridge Library Collection will bring back to life books of enduring scholarly value (including out-of-copyright works originally issued by other publishers) across a wide range of disciplines in the humanities and social sciences and in science and technology.

Essay on the Modifications of Clouds

Luke Howard

CAMBRIDGE
UNIVERSITY PRESS

CAMBRIDGE UNIVERSITY PRESS

Cambridge, New York, Melbourne, Madrid, Cape Town,
Singapore, São Paolo, Delhi, Tokyo, Mexico City

Published in the United States of America by Cambridge University Press, New York

www.cambridge.org
Information on this title: www.cambridge.org/9781108037686

© in this compilation Cambridge University Press 2011

This edition first published 1865
This digitally printed version 2011

ISBN 978-1-108-03768-6 Paperback

ESSAY

ON THE

MODIFICATIONS OF CLOUDS.

E.M. Williams lith.

M. & N Hanhart imp.

FORMS ASSUMED BY CLOUDS WHEN GATHERING FOR A THUNDERSTORM.

ESSAY

ON THE

MODIFICATIONS OF CLOUDS.

By LUKE HOWARD, F.R.S., &c.

[FIRST PUBLISHED 1803.]

THIRD EDITION.

LONDON:
JOHN CHURCHILL & SONS, NEW BURLINGTON STREET.
MDCCCLXV.

ADVERTISEMENT TO THE READER.

THIS Essay was one among several, on subjects connected with the study of the Climate of this country, which were presented by the Author to a Philosophical Society, of which he was a Member, and at its instance published. The following Note, attached by the Editor of the *Philosophical Magazine* to the first of these publications, will explain the nature of the Association here mentioned.

" Read before the *Askesian Society*, London:—This Society consists of a select number of gentlemen, associated for their mutual improvement in the different branches of Natural Philosophy. It was instituted in March, 1796, and the regular meetings are held every week during the winter."—*Philo. Mag.* vol. vii. p. 355.

During several years, while it continued in activity, the Author was a regular assistant at the meetings, and a contributor to the archives of this little Association; the proceedings of which, and the discussions arising out of questions moved by each member in his turn, were productive of no small intellectual benefit to the members.

The Essay on Clouds has been since, in various ways, reprinted and abridged, and is to be found in several Cyclopædias and periodical publications; but the author has not, until now, thought fit to replace the small, and long since dispersed, Edition of it, originally printed separately for his use.

London, 23d Fifth Month, 1832.

PREFACE TO THE THIRD EDITION.

It has been, for some years past, impossible to obtain the *Essay on the Modifications of the Clouds* in its separate form, the last edition, published in 1832, having been long exhausted. It appeared therefore desirable to bring out a new one. For, although the other scientific works by the same Author continue to be of value, this small treatise cannot but possess the highest interest of any, as it embodies that part of his labours which gave him so high a place among Meteorologists.

From the time when this nomenclature was first suggested (about 1803), it has been universally adopted by scientific men, and, indeed, by all writers.

The illustrations which appear with this edition are the result of an endeavour to reproduce a series of water-colour sketches from nature, made (with one exception) by the Author,* and annotated with the remarks which accompany them with the view of exemplifying some of the most usual forms of the Modifications.

---

* The landscapes were by Mr. Kenyon. No. 2 is not by the same hand, having been purchased as a good example of a Stratus cloud, to complete the set.

They therefore possess a peculiar interest, even though much more perfect representations are to be found elsewhere.*

We must, however, warn the young student of Meteorology (as the Author himself would have done, had he superintended this edition) against limiting his conceptions of the Modifications to the particular *forms* here represented; a correct comprehension of the subject is only to be obtained by a habitual observation of Nature—to quote the words of Goethe—"Wenn man die Lehre Howard's, beim Beobachten wohl nutzen will, so muss man die von ihm bezeichneten Unterschiede fest im Auge behalten, und sich nicht irre machen lassen wenn gewisse schwankende Erscheinungen vorkommen; man übe sich vielmehr dieselben auf die Hauptrubriken zurück zu führen."

"If any one wishes to make a good use of Howard's teaching to guide his observations, he must keep firmly fixed before his eyes the distinctions that the latter has laid down, and not allow himself to be led astray by the occurrence of certain indistinct appearances; but practise himself in referring these to the main rules under which they come."

This Essay at its first appearance attracted the notice of the poet above quoted, who took a great interest in the subject. His remarks on the system, and correspondence with the Author, are published† at some length in his works, together with a Poem which so graphically points out the peculiar beauties of each

---

* Two instances in very different styles of art will suffice, viz.: the beautifully engraved illustrations of Cirri, &c., by J. C. Armytage, in vol. v. of Ruskin's "Modern Painters," and the stereoscopic photographs of Mr. Wilson, of Aberdeen.

† Goethe's Sämmtliche Werke, vol. xl. p. 311, &c., Cotta's Edit. 1840. We may refer the reader who wishes for any biographical information respecting the Author to his letter to Goethe, in p. 342, &c.

class of clouds, that we think it will not be out of place to append it here, for the benefit of those who may not before have fallen in with it.

We refrain from inserting a translation, although some have been published, because the simple beauty of the original is for the most part completely lost in the attempt to render it in English verse.

W. DILLWORTH HOWARD.

ELIOT HOWARD.

Tottenham, November, 1864.

POEM ON THE CLOUDS,

BY GOETHE.

Wenn Gottheit Camarupa, hoch und hehr,

Durch Lüfte schwankend wandelt leicht und schwer,

Des Schleiers Falten sammelt, sie zerstreut,

Am Wechsel der Gestalten sich erfreut,

Jetzt starr sich hält, dann schwindet wie ein Traum—

Da staunen wir, und trau'n dem Auge kaum.

Nun regt sich kühn des eignen Bildens Kraft

Die Unbestimmtes zu Bestimmtem schafft;

Da droht ein Leu, dort wogt ein Elephant,

Kameeles Hals, zum Drachen umgewandt;

Ein Heer zieht an, doch triumphirt es nicht,

Da es die Macht am steilen Felsen bricht;

Der treuste Wolkenbote selbst zerstiebt,

Eh' er die Fern' erreicht, wohin man liebt.

Er aber, Howard, giebt mit reinem Sinn

Uns neuer Lehre herrlichsten Gewinn:

Was sich nicht halten, nicht erreichen lässt,

Er fasst es an, er hält zuerst es fest;

b 2

Bestimmt das Unbestimmte, schränkt es ein,
Benennt es treffend!—Sei die Ehre Dein!
Wie Streife steigt, sich ballt, zerflattert, fällt,
Erinnre dankbar Deiner sich die Welt.

STRATUS.

Wenn von dem stillen Wasserspiegel-Plan
Ein Nebel hebt den flachen Teppich an,
Der Mond, dem Wallen des Erscheins vereint,
Als ein Gespenst Gespenster bildend scheint,
Dann sind wir alle, das gestehn wir nur,
Erquickt', erfreute Kinder, o Natur!

Dann hebt sich's wohl am Berge, sammlend breit
An Streife Streifen, so umdüstert's weit
Die Mittelhöhe, beidem gleich geneigt,
Ob's fallend wässert, oder luftig steigt.

CUMULUS.

Und wenn darauf zu höh'rer Atmosphäre
Der tüchtige Gehalt berufen wäre,
Steht Wolke hoch, zum herrlichsten geballt,
Verkündet, festgebildet, Machtgewalt,
Und was Ihr fürchtet und auch wohl erlebt,
Wie's oben drohet, so es unten bebt.

CIRRUS.

Doch immer höher steigt der edle Drang!

Erlösung ist ein himmlisch leichter Zwang.

Ein Aufgehäuftes, flockig löst sich's auf,

Wie Schäflein trippelnd, leicht gekämmt zu Hauf—

So fliesst zuletzt was unten leicht entstand,

Dem Vater oben still in Schoos und Hand.

NIMBUS.

Nun lasst auch niederwärts, durch Erdgewalt

Herabgezogen was sich hoch geballt,

In Donnerwettern wüthend sich ergehn,

Heerschaaren gleich entrollen und verwehn!—

Der Erde thätig-leidendes Geschick!—

Doch mit dem Bilde hebet euren Blick:

Die Rede geht herab, denn sie beschreibt;

Der Geist will aufwärts, wo er ewig bleibt.

EXPLANATION OF PLATES.

Frontispiece.

A SKY full of specimens of the peculiar forms assumed by the clouds when gathering for a *Thunderstorm.* (The dense clouds in front, though characteristic as *forms*, are not enough shrouded in a gloomy distance.) The *Nimbus* behind, on the left, is better placed, and shows crossed, as is at times the case, by a dense *Cirro-stratus.* The *Cirro-cumulus*, next above, is highly characteristic in its structure; and the large, dense *Cirrus* above belongs equally to approaching *Thunder.*

" L. H.

" *March*, 1849."

No. 1.—OPPOSITE PAGE 5.

" In the blue sky, at top, the *Cirrus*, in some of its fair-weather shapes (but much too near the eye in position). The *feather* on the left is a rare, but genuine, variety of the cloud.

" Above the horizon are some large *Cumulus* clouds, receiving on their tops a subsiding *Cirro-stratus :* the manner in which these modifications unite upon occasion to form, ultimately, *Cumulo-stratus.*

" L. H.

" *March*, 1849."

No. 2.—OPPOSITE PAGE 7.

" This picture presents, mixed with the objects common to a landscape, the commencement of a *Stratus*, or evening mist, creeping, as it rises, through the valley, to become shortly a dense body of cloud, resting with a level surface on the ground like a lake of water; and possibly, on the morrow, covering the country with a *Fog.* Above, in the blue sky, are features of *Cirrus.*

" L. H.

" *March*, 1849."

No. 3.—Opposite Page 8.

"In the foreground of this Landscape by Kenyon (that is, of the *sky* part) is a *Cumulus* just breaking up in a fine summer's evening, the more distant spot to the left showing the little of the cloud that remained when a hasty sketch of it had been made. Next above, and actually at the greatest elevation of the three, is a *Cirrus;* at the top of the picture, a fine *Cirro-cumulus.*

"L. H.

"*March,* 1849."

No. 4.—Opposite Page 10.

"This is a fine specimen of the Modification Cumulo-stratus, illustrating the manner in which it is occasionally produced by *inosculation,* or the mutual attraction of a *Cumulus* and a *Cirro-stratus.* These elements of the compound appear also in the group *separate.*

"Above, in the blue sky, are Cirri tending to the change which makes them *Cirro-cumulus.*

"The Picture is by Kenyon; the sketch was by L. H., in a fine evening, after showers.

"L. H.

"*March,* 1849."

No. 5.—Opposite Page 12.

"A *Nimbus,* or Rain-cloud, letting fall a shower at a moderate distance: beneath and on the horizon are seen two clouds, as yet in the sunshine, which, by the expansion of their tops, show considerable forwardness of progress towards the same state.

"The Landscape is by Kenyon. The sky is one of a number of sketches given him by the author of the Modifications.

"L. H.

"*March,* 1849."

ESSAY,

ETC.

On the Modifications of Clouds, and on the Principles of their Production, Suspension, and Destruction; being the Substance of an Essay read before the Askesian Society in the Session 1802—3.

SINCE the increased attention which has been given to Meteorology, the study of the various appearances of water suspended in the Atmosphere is become an interesting and even necessary branch of that pursuit.

If Clouds were the mere result of the condensation of Vapour in the masses of atmosphere which they occupy, if their variations were produced by the movements of the atmosphere alone, then indeed might the study of them be deemed an useless pursuit of shadows, an attempt to describe forms which, being the sport of winds, must be ever varying, and therefore not to be defined.

But however the erroneous admission of this opinion may have operated to prevent attention to them, the case is not so with Clouds. They are subject to certain distinct modifications, produced by the general causes which effect all the variations of the Atmosphere: they are commonly as good visible indications of the operation of these causes as is the countenance of the state of a person's mind or body.

It is the frequent observation of the countenance of the sky, and of its connexion with the present and ensuing phænomena, that constitutes the ancient and popular Meteorology. The want of this branch of knowledge renders the predictions of the Philosopher (who in attending only to his instruments may be said to examine only the *pulse* of the atmosphere) less generally successful than those of the weatherwise Mariner or Husbandman.

With the latter, the dependence of their labours on the state of the atmosphere, and the direction of its Currents, creates a necessity of frequent observation, which in its turn produces experience.

But as this experience is usually consigned only to the memory of the possessor, in a confused mass of simple aphorisms, the skill resulting from it is in a manner incommunicable; for, however valuable these links when in connexion with the rest of the Chain, they often serve, when taken singly, only to mislead; and the power of connecting them, and of forming a judgment upon occasion from them, resides only in the mind before which their relations have passed, though perhaps imperceptibly, in review. In order to enable the Meteorologist to apply the key of Analysis to the experience of others, as well as to record his own with brevity and precision, it may perhaps be allowable to introduce a Methodical nomenclature, applicable to the various forms of suspended water, or, in other words, to the Modifications of *Cloud*.

By modification is to be understood simply the Structure or manner of aggregation, not the precise form or magnitude, which indeed varies every moment in most Clouds. The principal Modifications are commonly as distinguishable from each other as a Tree from a Hill, or the latter from a Lake; although Clouds in the same modification, considered with respect to each other, have often only the common resemblances which exist among trees, hills, or lakes, taken generally.

The Nomenclature is drawn from the Latin. The reasons for having recourse to a dead language for Terms to be adopted by the learned of different nations are obvious. If it should be asked why the *Greek* was not preferred, after the example of Chemistry, the author answers, that the objects being to be defined by visible characters, as in Natural history, it was desirable that the Terms should at once convey the idea of these, and render a frequent recourse to definitions needless to such as understand the literal sense, which many more would, it is concluded, in Latin than in Greek words.

There are three simple and distinct Modifications, in any one of which the aggregate of minute drops called a Cloud *may be formed, increase to its greatest extent, and finally decrease and disappear.*

But the same Aggregate which has been formed in one Modification, upon a change in the attendant circumstances may pass into another:

Or it may continue for a considerable time in an intermediate state, partaking of the characters of two Modifications; and it may also disappear in this stage, or return to the first Modification.

Lastly, aggregates separately formed in different modifications may unite and pass into one exhibiting different characters in different parts, or a portion of a simple Aggregate may pass into another modification without separating from the remainder of the mass.

Hence, together with the simple, it becomes necessary to admit intermediate and compound Modifications; and to impose names on such of them as are worthy of notice.

The simple Modifications are thus named and defined:

1. CIRRUS. *Def. Nubes cirrata, tenuissima, quæ undique crescat.*

Parallel, flexuous, or diverging fibres, extensible by increase in any or in all directions.

2. CUMULUS. *Def. Nubes cumulata, densa, sursum crescens.*

Convex or conical heaps, increasing upward from a horizontal base.

3. STRATUS. *Def.* *Nubes strata, aquæ modo expansa, deorsum crescens.*

A widely extended, continuous, horizontal sheet, increasing from below upward.*

The intermediate Modifications which require to be noticed are:

4. CIRRO-CUMULUS. *Def.* *Nubeculæ densiores subrotundæ et quasi in agmine appositæ.*

Small, well defined roundish masses, in close horizontal arrangement or contact.

5. CIRRO-STRATUS. *Def.* *Nubes extenuata subconcava vel undulata. Nubeculæ hujusmodi appositæ.*

Horizontal or slightly inclined masses attenuated towards a part or the whole of their circumference, bent downward, or undulated; separate, or in groups consisting of small clouds having these characters.

The compound Modifications are:

6. CUMULO-STRATUS. *Def.* *Nubes densa, basim planam undique supercrescens, vel cujus moles longinqua videtur partim plana partim cumulata.*

The Cirro-stratus blended with the Cumulus, and either appearing intermixed with the heaps of the latter *or superadding a wide-spread structure to its base.*

7. CUMULO-CIRRO-STRATUS *vel* NIMBUS. *Def.* *Nubes vel nubium congeries* [superné cirrata] *pluvium effundens.*

The Rain cloud. A cloud, or system of clouds from which rain is falling. It is a horizontal sheet, above which the Cirrus spreads, while the Cumulus enters it laterally and from beneath.

---

* This application of the Latin word *stratus* is a little forced. But the substantive *stratum*, did not agree in its termination with the other two, and is besides already used in a different sense even on this subject, e.g. *a stratum of clouds;* yet it was desirable to keep the derivation from the verb *sterno,* as its significations agree so well with the circumstances of this Cloud.

E.M.Williams lith.

M.& N.Hanhart.imp.

CUMULOSTRATUS FORMING. FINE WEATHER CIRRI ABOVE

Of the Cirrus.

Clouds in this Modification appear to have the least density, the greatest elevation, and the greatest variety of extent and direction. They are the earliest appearance after serene weather. They are first indicated by a few threads pencilled, as it were, on the sky. These increase in length, and new ones are in the mean time added to them. Often the first-formed threads serve as stems to support numerous branches, which in their turn give rise to others.

The increase is sometimes perfectly indeterminate, at others it has a very decided direction. Thus the first few Threads being once formed, the remainder shall be propagated in one or more directions laterally, or obliquely upward or downward,* the direction being often the same in a great number of Clouds visible at the same time; for the oblique descending tufts appear to converge towards a point in the Horizon, and the long straight streaks to meet in opposite points therein; which is the optical effect of parallel extension.

Their *duration* is uncertain, varying from a few minutes after the first appearance to an extent of many hours. It is long when they appear alone and at great heights, and shorter when they are formed lower and in the vicinity of other Clouds.

This Modification, although in appearance almost motionless, is intimately connected with the variable motions of the atmosphere Considering that Clouds of this kind have long been deemed a prognostic of wind, it is extraordinary that

---

* The upward direction of the fibres, or tufts of this Cloud is found to be a decided indication of the decomposition of vapour preceding *rain:* the downward as decidedly indicates *evaporation* and fair weather. In each case they point towards the place of the Electricity which is evolved at the time.

the nature of this connexion should not have been more studied; as the knowledge of it might have been productive of useful results.

In *fair* weather, with light variable breezes, the sky is seldom quite clear of small groups of the oblique Cirrus, which frequently come on from the leeward, and the direction of their increase is to windward. Continued *wet* weather is attended with horizontal sheets of this cloud, which subside quickly and pass into the Cirro-stratus.

Before *storms* they appear lower and denser, and usually in the quarter opposite to that from which the storm arises. Steady high winds are also preceded and attended by streaks running quite across the sky in the direction they blow in.

Of the Cumulus.

Clouds in this Modification are commonly of the most dense structure: they are formed in the lower atmosphere, and move along with the Current which is next the earth.

A small irregular spot first appears, and is, as it were, the *nucleus* on which they increase. The lower surface continues irregularly plane, while the upper rises into conical or hemispherical heaps; which may afterwards continue long nearly of the same bulk, or rapidly grow to the size of mountains.

In the former case they are usually numerous and near together, in the latter few and distant; but whether there are few or many, their *bases* lie always nearly in one horizontal plane; and their increase upward is somewhat proportionate to the extent of base, and nearly alike in many that appear at once.

Their appearance, increase, and disappearance, in *fair* weather, are often periodical, and keep pace with the Temperature of the day. Thus, they will begin to form some hours after sun-rise, arrive at their maximum in the hottest part of the afternoon, then go on diminishing, and totally disperse about sun-set.

E.M.Williams,lth.

STRATUS, OR GROUND FOG.

M.&N.Hanhart.imp.

But in *changeable* weather they partake of the vicissitudes of the atmosphere: sometimes evaporating almost as soon as formed; at others suddenly forming, and as quickly passing to the compound modifications.

The Cumulus of *fair* weather has a moderate elevation and extent, and a well-defined rounded surface. Previous to *rain* it increases more rapidly, appears lower in the atmosphere, and with its surface full of loose fleeces or protuberances.

The formation of large Cumuli to leeward in a strong wind, indicates the approach of a calm with rain. When they do not disappear or subside about sunset, but continue to rise, Thunder is to be expected in the night.

Independently of the beauty and magnificence it adds to the face of nature,* the Cumulus serves to screen the earth from the direct rays of the sun; by its multiplied reflections to diffuse, and, as it were, economise the Light, and also to convey the product of Evaporation to a distance from the place of its origin. The relations of the Cumulus with the state of the Barometer, &c., have not yet been enough attended to.

Of the Stratus.

This Modification has a mean degree of density.

It is the *lowest* of Clouds, since its inferior surface commonly rests on the earth or water.

Contrary to the last, which may be considered as belonging to the *day*, this is properly the cloud of *night*; the time of its first appearance being about sun-set. It comprehends all those creeping Mists which in calm evenings ascend in spreading

---

* The connexion of the finer rounded forms, and more pleasing dispositions and colours of these Aggregates, with warmth and calmness; and of everything that is dark and abrupt, and shaggy, and blotched, and horrid in them, with cold, and storm, and tempest, may be cited as no mean instance of the perfection of that Wisdom and Benevolence which formed and sustains them.

sheets (like an inundation) from the bottom of valleys, and the surface of lakes, rivers, and other pieces of water, to cover the surrounding country.

Its *duration* is frequently through the night.

On the return of the sun the level surface of this Cloud begins to put on the appearance of Cumulus, the whole at the same time separating from the ground. The continuity is next destroyed, and the Cloud ascends and evaporates, or passes off with the morning breeze. This change has been long experienced as a prognostic of fair weather,* and indeed there is none more serene than that which is ushered in by it.

Of the Cirro-Cumulus.

The Cirrus having continued for some time increasing or stationary, usually passes either to the Cirro-cumulus or the Cirro-stratus, at the same time descending to a lower station in the atmosphere.

The Cirro-cumulus is formed from a Cirrus, or from a number of small separate Cirri, by the fibres collapsing as it were, and passing into small roundish masses, in which the texture of the Cirrus is no longer discernible; although they still retain somewhat of the same relative arrangement. This change takes place either throughout the whole mass at once, or progressively from one extremity to the other. In either case the same effect is produced on a number of adjacent Cirri at the same time and in the same order. It appears in some instances to be accelerated by the approach of other Clouds.

This Modification forms a very beautiful sky, sometimes exhibiting numerous distinct beds of these small connected clouds, floating at different altitudes.

---

* At nebulæ magis ima petunt, campoque recumbunt.

VIRGIL, *Georg.* lib. i.

CUMULUS BREAKING UP, CIRRUS & CIRRO CUMULUS ABOVE.

The Cirro-cumulus is frequent in *summer*, and is attendant on warm and dry weather. It is also occasionally and more sparingly seen in the intervals of showers, and in *winter*. It may either evaporate, or pass to the Cirrus or Cirro-stratus.

Of the Cirro-Stratus.

This Cloud appears to result from the subsidence of the fibres of the Cirrus to a horizontal position, at the same time that they approach towards each other laterally. The form and relative position, when seen in the distance, frequently give the idea of shoals of fish. Yet in this, as in other instances, the *structure* must be attended to rather than the *form*, which varies much, presenting at times the appearance of parallel bars, or interwoven streaks like the grain of polished wood. It is thick in the middle, and extenuated towards the edge. The distinct appearance of a Cirrus, however, does not always precede the production of this and the last Modification.

The Cirro-stratus precedes *wind* and *rain*, the nearer or distant approach of which may sometimes be estimated from its greater or less abundance and permanence. It is almost always to be seen in the intervals of storms. Sometimes this and the Cirro-cumulus appear together in the sky, and even alternate with each other in the same cloud; when the different evolutions which ensue are a curious spectacle; and a judgment may be formed of the weather likely to ensue by observing which Modification prevails at last. The Cirro-stratus is the Modification which most frequently and completely exhibits the phenomena of the Solar and Lunar halo, and (as supposed from a few observations) the Parhelion and Paraselene also. Hence the reason of the prognostic of foul weather, commonly drawn from the appearance of Halo.*

---

* The frequent appearance of Halo in this cloud may be attributed to its possessing great extent, at such times, with little perpendicular depth, and the requisite degree of continuity of substance.

c

This Modification is on this account more peculiarly worthy of investigation.

Of the Cumulo-Stratus.

The different Modifications which have been treated of sometimes give place to each other, at other times two or more appear in the same sky, but in this case the Clouds in the same Modification lie mostly of the same plane, those which are more elevated appearing through the intervals of the lower, or the latter showing dark against the lighter ones above them. When the Cumulus increases rapidly, a Cirro-stratus is frequently seen to form around its summit, reposing thereon as on a mountain, while the former Cloud continues discernible in some degree through it. This state of things continues but a short time. The Cirro-stratus speedily becomes denser and spreads, while the superior Cumulus extends itself and passes into it, the base continuing as it was, while the convex protuberances change their position till they present themselves laterally and downward. More rarely, the Cumulus performs this evolution by itself, and its superior part then constitutes the incumbent Cirro-stratus.

In either case a large lofty dense Cloud is formed, which may be compared to a Mushroom with a very thick short stem. But when a whole sky is crowded with this Modification, the appearances are indistinct. The Cumulus rises through the interstices of the superior Cloud; and the whole, seen as it passes off in the distant horizon, presents to the fancy mountains covered with snow, intersected with darker ridges, lakes of water, rocks, and towers, &c. The *distinct* Cumulo-stratus is formed in the interval between the first appearance of the fleecy Cumulus and the commencement of rain, while the lower atmosphere is yet dry; also during the approach of Thunder-storms: the *indistinct* appearance of it is chiefly in the longer or shorter intervals of showers of rain, snow, or hail.

F.M.Williams lith.

M&N.Hanhart, imp.

CUMULOSTRATUS; AS PRODUCED BY THE INOSCULATION OF CUMULUS WITH CIRROSTRATUS.

CIRRI ABOVE, PASSING TO CIRROCUMULUS.

Of the Nimbus, or Cumulo-Cirro-Stratus.

Clouds in any one of the preceding Modifications, at the same degree of elevation, or in two or more of them at different elevations, may increase so as completely to obscure the sky; and may at times put on an appearance of density which to the inexperienced observer indicates the speedy commencement of *rain*. It is nevertheless extremely probable, as well from attentive observation as from a consideration of the several modes of their production, that *Clouds, while in any of these states, do not at any time let fall rain.*

Before this effect takes place they have been uniformly found to undergo a change, attended with appearances sufficiently remarkable to constitute a distinct Modification. These appearances, when the rain happens over-head, are but imperfectly seen. We can then only observe, before the arrival of the denser and lower Clouds, or through their interstices, that there exists *at a greater altitude* a thin light veil, or at least a hazy turbidness. When this has considerably increased, we see the lower Clouds spread themselves, till they unite in all points and form one uniform Sheet. The rain then commences; and the lower clouds, arriving from the windward, move under this Sheet and are successively lost in it. When the latter cease to arrive, or when the Sheet breaks, [letting through the sunbeams,] every one's experience teaches him to expect an abatement or cessation of *the rain.*

But there often follows, what seems hitherto to have been unnoticed, an immediate and great addition to the quantity of *cloud*. At the same time the actual obscurity is lessened, because the arrangement which now returns, gives freer passage to the rays of light: for on the cessation of rain, the lower broken clouds which remain rise into Cumuli, and the superior sheet puts on the various forms of the Cirro-stratus, sometimes passing to the Cirro-cumulus.

c 2

If the interval be long before the next shower, the Cumulo-stratus usually makes its appearance; which it also does sometimes very suddenly after the first cessation.

But we see the nature of this process more perfectly in viewing a distant shower in profile.

If the Cumulus be the only cloud present at such a time, we may observe its superior part to become tufted with nascent Cirri. Several adjacent Clouds also approach and unite laterally by subsidence.

The Cirri increase, extending themselves upward and laterally, after which the shower is seen to commence. At other times, the converse takes place of what had been described relative to the cessation of rain. The Cirro-stratus is previously formed above the Cumulus, and their sudden union is attended with the production of Cirri and rain.

In either case the Cirri *vegetate*, as it were, in proportion to the quantity of rain falling, and give the Cloud a character by which it is easily known at great distances, and to which, in the language of Meteorology, we may appropriate the nimbus of the Latins.*

When one of these arrives hastily *with the wind* it brings but little rain, and frequently some hail or driven snow.

In heavy showers, the central Sheet once formed, is, as it were, warped to windward, the Cirri being propagated above and against the lower current, while the Cumuli arriving with the latter are successively *brought to*, and contribute to reinforce it.

---

* Qualis ubi ad terras abrupto sidere *nimbus*
　It mare par medium, miseris heu *prescia longe*
　Horrescunt corda agricolis.—*Virgil.*

E.M. Williams lith.

M & N Hanhart imp.

NIMBUS, OR RAIN CLOUD.

Such are the phænomena of *showers*. In continued gentle rains it does not appear necessary for the resolution of the Clouds that the different Modifications should come into actual contact.

It is sufficient that there exist two strata of Clouds, one passing beneath the other, and each continually tending to horizontal uniform diffusion.* It will rain during this state of the two strata, although they should be separated by an interval of many hundred feet in elevation. See an instance in De Luc, *Idées sur la Météorologie*, tom. ii. p. 52, &c. [It is not to be supposed that the intermediate space is, on these occasions, at any time free from a conducting medium of diffused watery particles, enabling the opposite Electricities to neutralize each other.]

As the masses of Cloud are always blended, and their arrangements broken up before rain comes on, so the reappearance of these is the signal for its cessation. The thin sheets of Cloud which pass over during a wet day, certainly receive from the humid atmosphere a supply proportionate to their consumption, while the latter prevents their increase in bulk. Hence a seeming paradox, which yet accords strictly with observation, that for any given hour of a wet day, or any given day of a wet season, *the more cloud the less rain*. Hence also arise some further reflections on the purpose answered by Clouds in the Economy of nature. Since rain may be produced by, and continue to fall from, the slightest obscuration of the sky by the Nimbus (or by *two sheets* in different states,) while the Cumulus or Cumulostratus, with the most dark and threatening aspect, shall pass over without letting fall a drop, until the change of state commences, it should seem that the latter are Reservoirs [water-waggons they are called by some] in which the water is collected from a large space of atmosphere for occasional and local irrigation in dry seasons, and by means of which it is also arrested at times in its descent in the midst of

---

* The superior stratum is often seen, in this case, to partake of the Cirrus.

wet ones. In which so evident provision for the sustenance of all animal and vegetable life, as well as for the success of mankind in that pursuit so essential to their welfare, in temperate climates, of cultivating the earth, we may discover the wisdom and goodness of the Creator and Preserver of all things.

The Nimbus, although in itself one of the least beautiful Clouds, is yet now and then superbly decorated with its attendant the rainbow; which is seen in perfection when backed by the widely-extended uniform gloom of this Modification.

The relations of rain, and of periodical showers more especially, to the varying Temperature, Density, and Electricity of the atmosphere, will now probably obtain a fuller investigation, and with a better prospect of success, than heretofore.

As the establishing distinctive characters for Clouds has been heretofore deemed a desirable object, and it is consequently probable that the author's Modifications will begin to be noted in Meteorological registers as they occur, (a practice which may be productive of considerable advantage to science,) the following System of abbreviations may, perhaps, be found of some use in this respect. They will save room and the labour of writing, and types may be easily formed for printing them. These are advantages not to be despised, when observations are to be noted once or oftener in the day. It is only necessary that they be inserted in a column headed *Clouds;* that the Modifications which appear together be placed side by side, and those which succeed to each other in the order of the column, but separated by a line or space from the preceding and succeeding day's notations.

\ Cirrus :   ⌒ Cumulus :   — Stratus : \⌒ Cirro-cumulus : \＿ Cirro-stratus :   ⌒＿ Cumulo-stratus : \⌒＿ Cirro-cumulo-stratus, or Nimbus.

[In my first publication on Clouds, I was induced, by a supposed necessity arising from the novelty of the subject, to add to the definitions a set of plates, of the several modifications. I have now decided to omit these representations: being satisfied, both by reflection and experience, that the real student will acquire his knowledge in a more solid manner, by the observation of nature, without the aid of drawings, and that the more superficial are liable to be led into error by them.]*

In tracing the various appearances of clouds, we have only adverted to their connexion with the different states of the atmosphere, (on which, indeed, their diversity in a great measure depends,) having purposely avoided mixing difficult and doubtful explanation with a simple descriptive arrangement.

Of Evaporation.

On the remote and universal origin of clouds there can be but one opinion— that the water of which they consist has been carried into the atmosphere by Evaporation. It is on the nature of this process, the state in which the Vapour subsists for a time, and the means by which the Water becomes again visible, that the greatest diversity of opinion has prevailed.

The Chemical philosopher, seduced by analogy, and accustomed more to the action of liquids on solids, naturally regards Evaporation as a solution of water in the atmosphere, and the appearance of cloud as the first sign of its precipitation ; which becoming afterwards (under favourable circumstances) more abundant, pro- duces rain. The theory of Dr. Hutton goes a step further, assumes a certain rate of solution differing from that of the advance of temperature by which it is effected, and deduces a general explanation of clouds and rain from the precipitation which, according to his rule, should result from every mixture of different portions of saturated air. The fundamental principle of this theory has been disproved in

* The author did not remain exclusively of this opinion, as is shown in a subsequent work (*Seven Lectures on Meteorology*, p. 196, note *e*). The date of the annotations to the pictures now published also proves the same fact.—ED.

an essay heretofore presented to the Society,* and which was written under the opinion, at present generally adopted by chemists, that evaporation depends on a solvent power in the atmosphere, and follows the general rules of chemical solution.

The author has since espoused a theory of evaporation which altogether excludes the above-named opinion, (and consequently Dr. Hutton's also,) and considers himself in a great degree indebted to it for the origin of the explanation he is about to offer. It will be proper, therefore, to state the fundamental propositions of this theory, with such other parts as appear immediately necessary, referring for mathematical demonstrations and detail of experiments to the work itself, which is entitled "Experimental Essays on the Constitution of mixed Gases; on the Force of Steam or Vapour from Water and other Liquids in different Temperatures, both in a Torricellian Vacuum and in Air; on Evaporation; and on the Expansion of Elastic Fluids by Heat. By John Dalton."—See Memoirs of the Literary and Philosophical Society of Manchester, vol. v. part 2.—The propositions are as follows:

"1. When two elastic fluids, denoted by A and B, are mixed together, there is no mutual repulsion amongst their particles; that is, the particles of A do not repel those of B, as they do one another." Consequently, the pressure or whole weight upon any one particle arises solely from those of its own kind.

"2. The force of steam from all liquids is the same at equal distances above or below the several temperatures at which they boil in the open air: and that force is the same under any pressure of another elastic fluid as it is *in vacuo*. Thus the force of aqueous vapour of 212° is equal to 30 inches of mercury; at 30° below, or 182°, it is of half that force; and at 40° above, or 252°, it is of double the force. So likewise the vapour from Sulphuric ether, which boils at 102°, then supporting

---

* See Phil. Mag. vol. xiv. p. 55.

OF THE MODIFICATIONS OF CLOUDS.

30 inches of mercury, at 30° below that temperature has half the force, and at 40° above it double the force: and so in other liquids. Moreover, the force of aqueous vapour of 60° is nearly equal to half an inch of mercury when admitted into a Torricellian vacuum: and water of the same temperature, confined with perfectly dry air, increases the elasticity to just the same amount.

" 3. The quantity of any liquid evaporated in the open air is directly as the force of steam from such liquid at its temperature, all other circumstances being the same."

The following is part of the Essay on Evaporation.

" When a liquid is exposed to the air, it becomes gradually dissipated in it: the process by which this effect is produced we call *Evaporation*.

" Many Philosophers concur in the theory of chemical solution. Atmospheric air, it is said, has an affinity for water; it is a menstruum in which water is soluble to a certain degree. It is allowed notwithstanding by all, that each liquid is convertible into an elastic vapour *in vacuo*, which can subsist independently in any temperature. But as the utmost forces of these vapours are inferior to the pressure of the atmosphere in ordinary temperatures, they are supposed to be incapable of existing in it in the same way as they do in a Torricellian vacuum: hence the notion of affinity is induced. According to this theory of Evaporation, atmospheric air (and every other species of air for aught that appears) dissolves water, alcohol, ether, acids, and even metals. Water below 212° is chemically combined with the gases. Above 212° it assumes a new form, and becomes a distinct elastic fluid, called *steam*. Whether water first chemically combined with air, and then heated above 212°, is detached from the air or remains with it, the advocates of the theory have not determined. This theory has always been considered as complex, and attended with difficulties; so much so, that M. Pictet and others have rejected it, and adopted that which admits of distinct elastic vapours in

D

the atmosphere at all temperatures, uncombined with either of the principal constituent gases; as being much more simple and easy of explication than the other: though they do not remove the grand objection to it, arising from atmospheric pressure."

" On the Evaporation of Water below 212°.

" I have frequently tried the Evaporation at all the temperatures below 212°. It would be tedious to enter into a detail of all the experiments, but I shall give the results at some remarkable points.

" The evaporation from water of 180° was from 18 to 22 grains per minute, according to circumstances: or about one-half of that at 212°.

" At 164° it was about one-third of the quantity at the boiling temperature, or from 10 to 16 grains per minute.

" At 152° it was only one-fourth of that at boiling, or from 8 to 12 grains, according to circumstances.

" The temperature of 144° affords one-fifth of the effect at boiling; 138° gave one-sixth, &c.

" Having previously to these experiments determined the force of aqueous vapour at all the temperatures under 212°, I was naturally led to examine whether the quantity of water evaporated in a given time bore any proportion to the force of vapour of the same temperature, and was agreeably surprised to find that they exactly corresponded in every part of the thermometric scale: thus the forces of vapour at 212°, 180°, 164°, 152°, 144°, and 138°, are equal to 30, 15, 10, $7\frac{1}{2}$, 6, and 5 inches of mercury respectively: and the grains of water evaporated per minute in those temperatures were 30, 15, 10, $7\frac{1}{2}$, 6, and 5, also; or numbers proportional to these. Indeed it should be so, from the established law of mechanics that all effects are proportional to the causes producing them. The atmosphere, it should seem, obstructs the diffusion of vapour, which would otherwise be almost instantaneous, as in *vacuo;* but this obstruction is overcome in proportion to the force of the vapour. The obstruction, however, cannot arise from the weight of the atmosphere, as has till now been supposed; for then it would effectually prevent any vapour from rising under 212°; but it is caused by the *vis inertiæ* of the particles of air, and is similar to that which a stream of water meets with in descending amongst pebbles.

" The theory of evaporation being thus manifested from experiments in high temperatures, I found that if it was to be verified by experiments in low temperatures, regard must be had to the force of vapour actually existing in the atmosphere at the time. For instance, if water of 59° were the subject, the force of vapour of that temperature is 1-60th of the force at 212°, and one might expect the quantity of evaporation 1-60th also. But if it should happen, as it sometimes does in summer, that an aqueous atmosphere to that amount does already exist, the evaporation, instead of being 1-60th of that from boiling water, would be nothing at all. On the other hand, if the aqueous atmosphere were less than that, suppose one half of it, corresponding to 39° of heat, then the effective evaporating force would be 1-120th of that from boiling water. In short, the evaporating force must be universally equal to that of the temperature of the water, diminished by that already existing in the atmosphere.

In order to find the force of the aqueous atmosphere I usually take a tall cylindrical glass jar, dry on the outside, and fill it with cold spring water fresh from the well. If dew be immediately formed on the outside, I pour the water out, let it stand awhile to increase in heat, dry the outside of the glass well with a linen cloth, and then pour the water in again. This operation is to be continued till dew ceases to be formed, and then the temperature of the water must be observed; and opposite to it in the table will be found the force of vapour in the atmosphere. This must be done in the open air, or at a window; because the air within is generally more humid than that without. Spring water is generally about 50°, and will mostly answer the purpose the three hottest months in the year; in other seasons an artificial cold mixture is required. The accuracy of the result obtained this way, I think, scarcely needs to be insisted on. Glass, and all other hard, smooth substances I have tried, when cooled to a degree below what the surrounding aqueous vapour can support, cause it to be condensed on their surfaces into water. The degree of cold is usually from 1° to 10° below the mean heat of the twenty-four hours; in summer I have often observed the point as high as 58° or 59°, corresponding to half an inch of mercury in force; and once or twice have seen it at 62° In changeable and windy weather it is liable to a considerable fluctuation; but this is not the place to enlarge upon it.

"For the purpose of observing the Evaporation in atmospheric temperatures, I got two light tin vessels, the one six inches in diameter and half an inch deep, the other eight inches diameter and three-fourths of an inch deep, and made to be suspended from a balance. When any experiment, designed as a test of the theory, was made, a quantity of water was put into one of these, (generally the six-inch one, which I preferred,) the whole was weighed to a grain; then it was placed in an open window or other exposed situation for ten or fifteen minutes, and again weighed to ascertain the loss by evaporation: at the same time the temperature of the water was observed, the force of the aqueous atmosphere ascertained as above, and the strength of the current of air noticed. From a great variety of experiments made both in the winter and summer, and when the evaporating force was strong and weak, I have found the results entirely conformable with the above theory. The same quantity is evaporated with the same evaporating force thus determined, whatever be the temperature of the air, as near as can be judged; but with the same evaporating force, a strong wind will double the effect produced in a still atmosphere. Thus, if the aqueous atmosphere be correspondent to 40° of temperature and the air be 60°, the evaporation is the same as if the aqueous atmosphere were at 60° of temperature and the air 72°; and in a calm air the evaporation from a vessel of six inches in diameter, in such circumstances, would be about ·9 of a grain per minute, and about 1·8 grains per minute in a very strong wind; the different intermediate quantities being regulated solely by the force of the wind."

Of the Aqueous Atmosphere.

Having quoted so much of this essay as may suffice to exhibit the principles on which we shall proceed, it may be useful, before we do this, to recapitulate the following circumstances respecting the atmosphere of aqueous gas, or (for brevity) the Aqueous atmosphere.

1st. It is supplied by the process of Evaporation, which by this theory appears to be reduced to the immediate union of water with Caloric into a binary compound, *Aqueous gas.*

2ndly. The supply of vapour (by which term, for the purposes of Meteorology, we may denote aqueous gas) is regulated by the following circumstances:—1. Temperature of the evaporating water; being greater as this is higher, and *vice versá.* 2. Quantity of surface exposed. Since it is from the surface only of the mass that the vapour in common cases can escape, the supply is in direct proportion thereto. 3. Quantity of vapour already subsisting in the atmosphere: the evaporation being less (with an equal temperature and surface) in proportion as this is greater, and *vice versá.*

3rdly. The vapour thus thrown into the atmosphere is diffusible therein by its own elasticity: which suffices for its ascent to any height in a perfect calm. Yet, as in this case the *inertia* of the particles of air considerably resists its diffusion, so in the opposite one of a brisk current, the vapour, by the same rule, must in some measure be drawn along with the mass into which it enters.

4thly. The quantity of vapour which, under equal pressure, can subsist in a given mass of air, will be greater as the common temperature is higher, and *vice versá.**

Aqueous vapour is the only gas contained in the atmosphere which is subject to very sensible variations in quantity. These variations arise from its attraction for caloric being inferior to that of all the others. Hence, when a cold body, such as the glass of water in the experiment above quoted, is presented to the atmosphere, the other gases will only be cooled by it (and that at all known temperatures); but the vapour, after being more or less cooled, will begin to be decomposed, its caloric entering the body while the water is left on the surface.

The formation of Cloud is in all cases the *remote* consequence of a decomposition thus effected, the caloric escaping not into a solid or liquid, but into the surrounding gases.

Of the Formation of Dew.

Dew is the *immediate* result of this decomposition. The particles of water constituting it are, singly, invisible, on account of their extreme minuteness.

---

* " The aqueous vapour atmosphere is variable in quantity according to temperature; in the torrid zone its pressure on the surface of the earth is equal to the force of ·6, and from that to one inch of mercury. In these parts it rarely amounts to the pressure of ·6, but I have frequently observed it above half an inch in summer; in winter it is sometimes so low as to be of no more force than 1 of an inch of mercury, or even half a tenth, in this latitude, and consequently much less where the cold is more severe. This want of equilibrium in the aqueous vapour atmosphere is a principal cause of that constant inundation of it into the temperate and frigid zones, where it becomes in part condensed in its progress by the cold (like the vapour of distillation in the worm of a refrigeratory), and supplies the earth with rain and dew."—See the Essays above quoted.

The approach of dew is, nevertheless, discoverable by a dark hazy appearance, verging from purple to faint red, extending from the horizon to a small distance upward, and most conspicuous over valleys and large pieces of water.

The theory of dew seems to be simply this :—During the heat of the day a great quantity of vapour is thrown into the atmosphere from the surface of the earth and waters. When the evening returns, if the vapour has not been carried off in part by currents, it will often happen that more remains diffused in the general atmosphere than the temperature of the night will permit to subsist under the full pressure of the aqueous atmosphere. A decomposition of the latter then commences, and is continued until the general temperature and aqueous pressure arrive at an equilibrium, or until the returning sun puts an end to the process. The caloric of the decomposed vapour goes to maintain the general temperature; while the water is separated in drops; which, minute as they are, arrive successively at the earth in the space of a few hours. That the ordinary production of dew is by a *real* descent of water from the atmosphere, and not by decomposition of vapour on surfaces previously cooled, (as in the experiment already mentioned,) any one may readily be convinced by observing in what abundance it is collected by substances which are wholly unfit to carry off the requisite quantity of caloric for the latter effect.

Of the Formation of the Stratus.

The case which has been just stated, *of the decomposition of vapour by the atmosphere in which it is already diffused,* goes but a little way in explanation of the production of a Cloud consisting of visible drops, and confined to a certain space in the atmosphere : much less does it enable us to account for the diversity of its situations and appearances. In attempting this we will begin with the Stratus, as the most simple in structure, and the next step, as it were, in the progress of *nubification.*

When dew falls upon a surface the temperature of which is superior to that of the atmosphere, it is plain that it will not continue there, but will be evaporated again: and a body so circumstanced will continue to refund into the atmosphere the whole of the water thus *gradually* deposited on it, so long as its substance can supply the requisite temperature to the surface. Moreover, water, either in mass or diffused among sand, clay, vegetable earth, &c., will continue to be evaporated therefrom with a force proportioned to its temperature, so long as the latter continues above that point which counterbalances the pressure of the Aqueous atmosphere.

From these causes it happens, that after the earth has been superficially dried by a continuance of sunshine, and heated, together with the lakes and rivers, to a considerable depth, there is an almost continual emission of vapour into the atmosphere by night.

This nocturnal evaporation is usually most powerful in the autumn, about the time that the temperature of the nights undergoes a considerable and some-times pretty sudden depression, attended with a calm.

In this state of things the vapour arising from the heated earth is condensed *in the act of diffusing itself*: the cold particles of water thus formed, in *descending*, meet the ascending stream of vapour, and condense a portion on their surfaces. If they touch the earth they are again evaporated, which is not necessarily the case if they alight on the herbage.* In this way an aggregate of visible drops is sooner or later formed: and as from the temperature thus communicated to the air next the earth, the vapour has still further and further to rise in order to be condensed, the cloud will be propagated upward in proportion.

Hence the Stratus most usually makes its appearance in the evening succeed-

---

* A plentiful dew may often be found on the grass after a Stratus.

ing a clear warm day, and in that quiescent state of the atmosphere which attends a succession of these. Hence also the frequency of it during the penetration of the autumnal rains into the earth; while in Spring, when the latter is *acquiring* temperature together with the atmosphere, it is [more] rarely seen.

Of the Formation of the Cumulus.

When the sun's rays traverse a clear space of atmosphere, it is well known that they communicate no sensible increase of temperature thereto. It is by the contact, and what may be termed the *radiation*, of opaque substances exposed to the light, that Caloric is thrown into the atmosphere.

This effect is first produced on the air adjacent to the earth's surface; and proceeds upward, more or less rapidly, according to the season and other attendant circumstances. In the morning, therefore, Evaporation usually prevails again; and the vapour, which continues to be thrown into air now increasing in temperature, is no longer condensed. On the contrary, it exerts its elastic force on that which the nocturnal temperature had not been able to decompose, and which consequently remained universally diffused. The latter, in rising *through the atmosphere* to give place to the supply from below, must necessarily change its climate, quit the lower air of equal temperature, and arrive among more elevated and colder air; the pressure from above still continuing unabated. The consequence is a partial decomposition, extending through the portion thus thrown up, and, in short, a recommencement in the superior region, of the same process which in the vicinity of the earth furnished the dew of the night. In this case, however, the particles of water cannot arrive at the earth, as they are necessarily evaporated again in their descent.

It appears that this second Evaporation takes place at that elevation where the temperature derived from the action of the sun's rays upon the earth, and de-

E

creasing upward, becomes just sufficient to counterbalance the pressure of the superior vapour.

Here is formed a sort of boundary between the region of cloud and the region of permanent vapour, which for the present purpose, and until we are furnished with a nomenclature for the whole science of Meteorology, may be denominated the Vapour plane.

Immediately above the Vapour plane, then, the formation of the Cumulus commences (as soon as a sufficient quantity of vapour has been thrown up) by the mixture of descending minute drops of water with vapour newly formed and just diffusing itself, as in the case of the Stratus before described.

A continuance of this process might be expected to produce a uniform sheet of cloud; in short, a Stratus, only differing in situation from the true one. Instead of which we see the first-formed small masses become so many centres, towards which all the water afterwards precipitated appears to be attracted from the space surrounding them; and this attraction becomes more powerful as the cloud increases in magnitude, insomuch that the small clouds previously formed disappear when a large one approaches them in its increase, and seem to vanish instead of joining it. This is probably owing to the small drops composing them having passed in a loose manner and successively, by attraction, into the large one.

Are the distinct masses into which the drops form themselves, in this case, due to the attraction of aggregation alone, or is the operation of any other cause to be admitted?

A rigid mathematician would perhaps answer the latter clause in the negative; and with such a conclusion we should have great reason to remain satisfied, as cutting short much of the inquiry that is to follow, were it not that it leaves us quite in the dark, both as to the cause of the variety so readily observable in clouds, and that of their long suspension, not to insist on several facts contained in the former part of this paper, which would then remain unaccounted for.

The operation of one simple principle would produce an effect at all times *uniform*, and varying only in degree. We should then see no diversity in clouds but in their magnitude; and the same attraction that could bring minute drops of water together through a considerable space of atmosphere in a few minutes, ought not to end there, but to effect their perfect union into larger, and finally into rain.

In admitting the constant operation of Electricity, which is at times so manifestly accumulated in clouds, upon their forms and arrangements, we shall not much overstep the limits of experimental inquiry, since it has been ascertained by several eminent philosophers, that " clouds, as well as rain, snow, and hail, that fall from them, are almost always electrified."*

An insulated Conductor formed of solid matter retains the charge given to it so much the longer, as it is more nearly spherical, and free from points and projecting parts. The particles of water, when charged, appear to make an effort to separate from each other, or, in other words, become mutually repulsive. Moreover, when a small conducting substance is brought within the reach of a large one similarly electrified, the latter, instead of repelling, will throw the small one into an opposite state, and then attract it. From these and other well-known facts in Electricity it would not be difficult to show, that an assemblage of particles of water floating in the atmosphere and similarly electrified, ought to arrange themselves in a spherical aggregate, into which all the surrounding particles of water (within a certain distance) should be attracted; at the same

* Cavallo. Complete Treatise on Electricity, vol. i. p. 74.

time the drops composing such aggregate should be absolutely prevented from uniting with each other *during the equilibrium of their electricity.*

To apply this reasoning to the formation of the Cumulus, we may, in the first place, admit that the commencement of distinct aggregation, in the descending particles of water, is due to their mutual attraction; by virtue of which small bodies, floating in any medium, tend to coalesce. The masses thus formed, however, often increase more rapidly than could be expected from the effect of simple attraction exercised at great distances. And when the cloud has arrived at a considerable size, its protuberances are seen to form, and successively sink down into the mass, in a manner which forces one to suppose a shower of invisible drops rushing upon it from all parts.

In unsettled weather the rapid formation of large Cumuli has been observed to clear the sky of a considerable hazy whiteness; which on the other hand has been found to ensue upon their *dispersion.**

On these considerations we are obliged to admit as a co-operating cause of the increase of this cloud, that sort of attraction which large insulated conducting masses exercise, when charged, on the smaller ones which lie within their influence. Instead of a *spherical* aggregate, however, we have only a sort of hemisphere; because that part of the cloud which presents itself toward the earth can receive no addition from beneath; there being in that direction no condensed water. On the contrary, the mass must be continually suffering a diminution

---

* That clouds are not always *evaporated* when they disappear, but sometimes dispersed so as to become invisible as distinct aggregates, is a fact pretty well ascertained by observation. This happens sometimes by the approach of other clouds; at others, the evaporation of part of a Cumulus is followed by the dispersion of the remainder. The criterion used was the speedy production of transparency in the one case, and of hazy turbidness in the other.

there, by the tendency of the cloud to subside and of the vapour plane to rise, during the increase of the diurnal temperature. It is this evaporation that cuts off all the Cumuli visible at one time in the same plane; and it is reasonable to conclude that much of the vapour thus produced is again condensed without quitting the cloud, as its course would naturally be mostly upward. Thus the drops of which a Cumulus consists may become larger the longer it is suspended, and the electricity stronger from the comparative diminution of surface.

Such is probably the manner in which this curious structure is raised, while the base is continually escaping from beneath it. That we may not, however, be accused of building a castle in the air by attempting further conjectures, we may leave the present Modification, after recapitulating some of its circumstances which appear to be accounted for.

The Cumulus is formed only in the daytime, because the direct action of the sun's rays upon the earth can alone put the atmosphere into that state of inequality of temperature which has been described. It evaporates in the evening from the cessation of this inequality, the superior atmosphere having become warmer, the inferior colder, attended with a decrease of the superficial Evaporation. It begins to form some hours after sunrise, because the vapour requires that space of time to become elevated by the gradual accession from below. When a *Stratus* covers the ground at sunrise, however, we often see it collect into Cumuli upon the Evaporation of that part of it which is immediately contiguous to the earth. And this ought to happen; for the Cloud is then insulated, the vapour plane is established, and everything in the same state (except in point of elevation) as in the ordinary mode of production of the Cumulus.

Lastly, the Cumulus, however dense it becomes, does not afford Rain, because it consists of drops similarly electrified and repelling each other; and is moreover continually evaporating from the plane of its Base. The change of form which comes on before it falls in Rain, and which indicates a disturbance of its Electrical state, will be noticed hereafter.

Of the Formation of the Cirrus.

It must have been owing entirely to the want of distinctive characters for clouds, and the consequent neglect of observing their changes, that the nature of this Modification more especially has not engaged the attention of Electricians. The attraction of aggregation operating on solid particles diffused in fluids, does indeed produce a great variety of ramifications in the process of crystallization: but these are either uniform in each substance, or have a limited number of changes. And in no instance do we see the same substance, separated from the same medium and unconfined in its movements, rival the numerous metamorphoses of the Cirrus.

The great elevation of these clouds in their ordinary mode of appearance has been ascertained both by geometrical observations,* and by viewing them from the summits of the highest mountains, when they appear as if seen from the plain. A more easy and not less convincing proof may be had by noting the time during which they continue to reflect the different coloured rays after sunset; which they do incomparably longer than any others. The same configuration of Cirrus has been observed in the same quarter of the sky for two successive days, during which a smart breeze from the opposite quarter prevailed below.

It is therefore probable that this Modification collects its water in a comparatively calm region; which is sometimes incumbent on the current next the earth,

---

* "The small white streaks of condensed vapour which appear on the face of the sky in serene weather, I have, by several careful observations, found to be from three to five miles above the earth's surface."—DALTON.

and almost out of the reach of its *daily* variations in temperature and quantity of vapour; but at other times is interposed between the latter and a supervening current from another climate: in which case it may be affected by both currents.

The Cumulus has been just now considered as an insulated body, consisting of moveable parts which accommodate themselves to the state of a *retained* Electricity. We shall attempt to explain the nature of the Cirrus by comparing it to those imperfect conductors, which being interposed between Electrics and Conductors, or between the latter in different states, serve to restore by degrees the equilibrium of the Electric fluid.

If a lock of hair be properly fixed on the prime conductor and electrified *plus*, the hairs will be separately extended at as great a distance from each other as possible; in which state they will continue for some time. The reason appears to be, that the contiguous air is then *minus;* and consequently these two moveable substances put themselves into the state most favourable to a communication which is going on slowly between bad conductors.

The same appearances will take place if the lock be electrified *minus*, the contiguous air being *plus;* and in each case the hairs will move *from* a body similarly electrified and brought near them, and *towards* one contrariwise electrified. Moreover, if we could insulate such a charged lock in the midst of a perfectly tranquil atmosphere of sufficient extent, in which particles of conducting matter were suspended, it is plain the latter would be attracted by it so long as the charge continued; after which they would be at large as before.

Dry air being an electric, and moist air but an indifferent conductor, it is reasonable to suppose that an immediate communication of Electricity between masses of air differently charged can scarcely happen to any great extent, except by the intimate mixture of such masses; an effect which may possibly follow in some cases, and occasion strong winds and commotions in the atmosphere. If we

consider how frequently, and to what an extent, the Electricity of the air is dis·
turbed (as appears from numerous experiments) by evaporation, by the formation
and passage of clouds, by elevation or depression of temperature, (by friction upon
surfaces of ice?) it seems probable that the particles of water floating in a calm
space may be frequently converted into conductors; by which the equilibrium is in
part restored after such disturbance.

Viewing the Cirrus in this light, it becomes important for those who are well
versed in electricity to study its appearances, and compare them with the changes
that ensue in the atmosphere.   A number of observations, made hitherto chiefly in
one place, and without system or aid from concurrent ones in others, have furnished
the preceding data, which may serve as hints for future investigation.

At present we can only conjecture that the local detached Cirri which ramify
in all directions, are collecting particles of water from the surrounding space; and
at the same time equalizing their own electricity with that of the air or vapour.

That when numerous oblique short tufts appear, they are conducting between
the air above and that below them.

That a decided direction of the extremities of pendent or erected Cirri from
the mass they join towards any quarter, is occasioned by the different Electricity of
a current of air whch is pressing upon the space they are contained in.   This is
the most important point to attend to, as these *tails* sometimes veer half round the
compass in the course of a few hours: and many observations have confirmed the
fact that they point *towards the coming wind*, and are larger and lower as this is
about to be stronger.

Lastly, the Cirri in parallel lines, stretching from horizon to horizon, denote a
communication of Electricity carried on through these clouds *over* the place of
observation; the two predisposed masses of atmosphere being very distant, and the
intermediate lower atmosphere not in a state to conduct it.   It is at least a circum-

stance well deserving inquiry, by what means the clouds in stormy seasons become arranged in these elevated parallel bars; which must be at least sixty miles long, and are probably much more, considering their elevation and that both extremities are often invisible.

Of the Nature of the Intermediate Modifications.

The conversion of the Cirrus into the Cirro-cumulus is a phenomenon which at some seasons may be daily traced, and which serves to confirm the opinion that there exists somewhat of the same difference between the Cumulus and the Cirrus, as between a *charged* and a *transmitting*, or an influenced conductor, among solid bodies. On this supposition, the orbicular arrangement of the particles ought to take place as soon as the mass has ceased to conduct from particle to particle, or to be so acted on by a contiguous conductor, as to have a *plus* and *minus* state within itself. And as this sort of communication in a cloud may be as slow as in other imperfect conductors, the equilibrium among the particles may be restored at one extremity some time before the other has ceased to transmit; whence a visible progress of the change, which may be traced in a Cirrus of sufficient length.

That an extensive horizontal Cirrus should become divided across its length, and that these divided parts should assume more or less of a round form, is also consistent with the idea of a change of this sort.* It is not so easy to give a reason why these small orbicular masses should remain in close arrangement, or even in contact, for several hours, forming a system of small clouds which yet

---

* A quickly evaporating Cumulus sometimes leaves a regular Cirrus behind, formed out of the remnant of the cloud, which, in the intermediate state, and just when it begins to show the sky through it, exactly represents the pores and fibre of *sponge*. [What is also curious, this appearance is a decided indication of drought approaching.]

F

do not interfere with each other or run together into one, but remain as it were in readiness to re-form the Cirrus; which sometimes happens very suddenly, though they more frequently evaporate by degrees.

The same remark applies to the curious, and as it were capricious divisions and subdivisions, both longitudinal and transverse, which happen in the Cirro-stratus when this cloud is verging towards the Cirro-cumulus. In general, nevertheless, its appearance is sufficiently distinct from that of the Cirrus and Cirro-cumulus. The Cirrus by the great extent in proportion to its mass, its distinct lines and angular flexures, and the Cirro-cumulus by the roundness and softness of its forms, indicates an essential difference in the state of the containing atmosphere. The Cirro-stratus appears to be always in a subsiding state, and to be more feebly acted on by Electricity than the preceding Modifications. Indeed, the lower atmosphere is usually pretty much charged with dew or haze at the time of its appearance, and therefore in a state to conduct a charge to the earth.

Of the Nature of the Compound Modifications, and of the Resolution of
Clouds into Rain.

From the theory of Evaporation it appears that no permanent cloud can be formed in the atmosphere, however low the temperature, without a sufficient pressure from vapour previously diffused. Hence, although in cold weather the breath and perspiration of animals, as also water at a certain excess of temperature, occasion a visible cloud, yet this cloud speedily evaporates again at all times, except when precipitation is actually going on at large in the atmosphere next the ground; when it is only *dispersed* therein. By comparing the different effects of a clear frosty air, and of a misty though much warmer one, on the perspiration and breath of horses warmed by labour, we may be assisted in reasoning on the great case of Evaporation, which, in some sense, is the *perspiration* of the earth.

The most powerful predisposing cause of Evaporation appears to be a superior current in the atmosphere, coming from a region where the low temperature of the surface, or its dry state, occasions a comparative deficiency of vapour. Hence, after heavy rain in winter, we see the sudden Evaporation, first of the remaining clouds, then of the water on the ground, followed by a brisk Northerly wind and sharp frost.

The very snow which had fallen on its arrival sometimes totally evaporates during the prevalence of such a wind. On the contrary, the first appearance of clouds forming in cold weather gives us to expect a speedy remission of the frost, although the cause is not generally known to be a change to a Southerly direction already begun in the superior atmosphere; which consequently brings on an excess of vapour.

This excess of vapour coming with a superior current, may be placed next to depression of temperature among the causes of Rain. The simultaneous decomposition of the higher *imported* vapour, and of that which is formed on the spot, or already diffused in the inferior current, would necessarily produce two orders of cloud; differing more or less in Electricity as well as in other respects. To the slow action of these upon each other, while Evaporation continues below, may be attributed the singular union which constitutes the Cumulo-stratus. It is too early to attempt to define the precise mode of this action, or to say by what change of state a Cumulus already formed is thrown into this Modification. That the latter phenomenon is an Electrical effect, no one who has had an opportunity to see its rapid progress during the approach of a thunderstorm can reasonably doubt.

To assert that rain is in almost every instance the result of the Electrical action of clouds upon each other, might appear to many too speculative, were we even to bring the authority of Kirwan for it, which is decidedly in favour of this idea of the process; yet it is in a great measure confirmed by observations made in various ways upon the Electrical state of clouds and of rain—not to insist on the probability that a thunder-storm is only a more sudden and sensible display of those energies which, according to the order observable in the Creation in other respects, ought to be incessantly and silently operating for general and beneficial purposes.

In the formation of the Nimbus, two circumstances claim particular attention : the spreading of the superior masses of cloud in all directions, until they become, like the Stratus, one uniform sheet; and the rapid motion and visible decrease of the Cumulus when brought under the latter. The Cirri, also, which so frequently stretch from the superior sheet upward, and resemble erected hairs, carry so much the appearance of temporary conductors of Electricity extricated by the sudden union of its minute drops into the vastly larger ones which form the rain, that one is in a manner compelled, when viewing this phenomenon, to indulge a little in Electrical speculations. By one experiment of Cavallo's, with a kite carrying three hundred and sixty feet of conducting string, in an interval between two showers, and kept up during rain, it seems that the superior clouds possessed a positive Electricity *before* the rain, which on the arrival of a large Cumulus gave place to a very strong negative, continuing as long as it was over the kite. We are not, however, warranted from this to conclude the Cumulus which brings on rain to be always negative; as the same effect might ensue from a positive Cumulus uniting with a negative Stratus. Yet the general negative state of the lower *atmosphere* during rain, and the positive indications commonly given by the true Stratus, render this the more probable opinion. It is not, however, absolutely necessary to determine this, seeing there is sufficient evidence in favour of the conclusion, that clouds formed in different parts of the atmosphere operate on each other, when brought near, occasioning their destruction by each other; an effect which can only be attributed to their possessing beforehand, or acquiring at the moment, the opposite Electricities.

It may be objected that this explanation is better suited to the case of a shower than to that of continued rain, for which it does not seem sufficient. If it should appear, nevertheless, that the supply of each kind of cloud may be kept up in proportion to the consumption, the objection will be answered. Now it is a well-known fact, that Evaporation from the surface of the earth and water returns and continues during rain, and consequently affords the lower clouds, while the upper may be recruited from vapour brought by the superior current, and continually subsiding in the form of dew; as is evident both from the turbidness of the atmosphere in rainy seasons, and from the plentiful deposition of dew in the nocturnal intervals of rain. Neither is it pretended that Electricity is any further concerned in the production of rain than as a secondary agent, which modifies the effect of the two grand predisposing causes—*a falling temperature and the influx of vapour.*

THE END.

LONDON :
SAVILL AND EDWARDS, PRINTERS, CHANDOS STREET,
COVENT GARDEN.

Printed in the United States
By Bookmasters